I Bet
Your House
Is Spotless

I Bet
Your House
Is Spotless

*Memories of working with people with OCD
and other mental health issues*

Matt Birks

PublishNation
www.publishnation.co.uk

Foreword

In writing this book I wanted to acknowledge the bravery of the people who allowed me to try to help them (with varying degrees of success), the colleagues who have inspired and supported me, and my wife Jeanette who has been there since before the start of my voyage into the world of mental health nursing but always there to pick up the wreckage I came home with and probably still do to this day. Heaven knows where I would be without you now.

Ramblings loosely about the author

I came to train as a mental health nurse in my mid-20s after deciding I ought to grow up and get a mortgage. This was the late 1980s and jobs were scarce – good ones more so – a bit like today. I had had a fairly laid-back early adulthood doing various dead-end jobs and had seen a bit of life and the world, so perfect qualifications for mental health nursing. The year before I started training, I got a job in a nursing home caring for older adults with dementia. I thought it would help, to have relevant experience. The hard, often smelly work for little money and working shifts had its lighter moments but was a world away from some of the later jobs I had. Nursing is like that, the huge variety – of settings and people is like few other jobs. I often hated it and many times swore I'd pack it in. Now of course I'm glad I didn't. Not that I could do it now. Years of political indifference towards mental health, despite what we hear the politicians promising, have all but destroyed the services such as day hospitals and many community initiatives. In less than 30 years we have moved from a situation where students were paid employees through having a bursary to having to pay thousands for the privilege of training while working much longer hours than traditional undergraduates. The pressure on services too has never been higher and I am full of admiration for those who still manage to make lives better under such ridiculous constraints. If I had my time again, but under today's circumstances, I might have done something else – be a used car salesman or something. Thank goodness not everyone is like me.

Since 2004 I have taught in universities training many, many mental health nurses of today. Some of these former students are teachers with students of their own now and some have gone abroad to work. I still work part-time as a therapist but in 2010 I spent a year working in the private sector – an experience I will never repeat and no exaggeration to say the worst year of my life.

I will be retired from clinical practice by the time you read this. I hope you get something from reading these tales. There have been some fascinating people along the way - some were genuinely warm kind people, most were service users, a few were colleagues. Some colleagues were generous and warm, one or two should never have been allowed within a mile of anyone vulnerable. I did whistle-blow once – it cost me a great deal of stress and abuse for a long time but enough said about that. Overall, I hope I did more good than harm.

Introduction

The role of the mental health nurse (MHN) has changed radically over the last 100 years or so. During that time, the primary aim has evolved from one of custodian to that of enabler. When hospital-based care was the norm, the nurse had the role of guardian. Patients – for that is what people in their care were called then, were cared for. Patients were fed, medicated, observed, given a place to stay in the true sense of the word asylum. Industrial therapy was the norm and patients were discouraged from leaving the hospital grounds until such a time as they were deemed 'cured' - which for many, was never.

Mental health being what it is – a socially defined label, (Rogers & Pilgrim, 2005), patients were at times admitted for what nowadays would seem to be ridiculous reasons, such as being an unmarried mother (Nolan, 2000). Henceforth they were disempowered as their independence was first physically and then psychologically removed in the name of care – many spent the rest of their lives within the institution.

This was not, on the whole, an act of malice. For many, this was an act of compassion. They were cared for in hospitals which, to modern eyes, seem dark and forbidding but it cannot be denied, were often set in beautiful grounds in quiet and peaceful locations – a far cry from some facilities today.

The role of the nurse was much simpler. Some may say less skilled but this is perhaps unfair. For what is the primary role of the mental health nurse? For Foster & Bennett, (2001), the primary goal is to facilitate the holistic recovery of those in their care. It is difficult to conceive of this being achieved without the ability to form and sustain a therapeutic relationship. The therapeutic relationship, also sometimes called the therapeutic alliance, (Rauch *et al.* 2012) refers to the relationship between the healthcare professional and a client. It is the means by which a therapist and client hope to engage with each other and effect beneficial change with the client (Barker, 2008). This ability is an often overlooked but pivotal skill (Norman & Ryrie 2013) and is the cornerstone of mental health nursing practice. It wouldn't make for exciting television in the way A&E based hospital dramas do but it makes all other interventions of worth possible for it generates trust.

When I first ventured into mental health nursing as a student in 1988, listening skills and the therapeutic relationship were the most valued skills. Even knowledge of the disease process was pushed to the background. Indeed, the anti-psychiatry movement was still influential and diagnosis was considered to be labelling and labelling was wrong, therefore the medical model was strongly discouraged. So, when we ventured onto the wards, armed only with an agreeable disposition and a willingness to engage, we learned about working with people first and patients second. We learned about their experiences with life, illness and the medications they took, primarily from our interactions with them. Of course, we also worked closely with doctors and other professionals from whom we learned about the MHA and medications from a prescriber's perspective but these were considered to be adjuncts to our essential 'people skills'. However, after I had been qualified for a few years, it became apparent to me that valuable as these skills were, they would only go so far. Working in an adult acute community day hospital setting, I felt I needed to be able to offer something more practical than what was starting to feel like little more than good intentions.

I was struck, as a student nurse, by how many service users returned to hospital on a regular basis. In my naivety, I had assumed people entered hospital, we cured and discharged. I soon learned this was not the case at all. Mental health recovery is more nuanced and now recognised as a more realistic goal (Andressen *et al.* 2011). In my younger days, I would have considered this lacking ambition. Now much older and hopefully a little wiser, I have modified my views. Still, it was my introduction to CBT on my elective placement in year three of my training that I witnessed clients (as they are usually called in this type of setting) making huge amounts of progress and getting back to their normal lives remarkably quickly. This was at a psychotherapy unit and from then on, because of its obvious efficacy, I was smitten with the CBT bug. Yes, it can be criticised. No, it doesn't suit everyone. It doesn't delve deep enough into the past for some. It can seem too formulaic – especially when practiced by inexperienced therapists. It also has to be said that the team were very selective about whom they thought was 'suitable' for therapy and who wasn't. However, for me, it was the first thing I had seen real progress in helping people with some very debilitating and distressing problems, relatively quickly and with minimal medication

(medication was seen as a barrier to progress for many with anxiety disorders). They metaphorical cherry though was that nurses ran the service, not psychiatrists. This was brilliantly empowering and for the first time in three years, I was sure I had made the right choice in mental health nursing. Here was something I could do that would give me more than satisfaction and validation - a feeling that I was making a measurable difference – and in CBT, quantifying progress is all. Just don't let this put you off though. Newer, '3rd wave' developments such as compassion-focused therapies use elements of CBT in a more individualised way. It is to this future that we should look to answer many of the criticisms of CBT and in particular the work of Professor Paul Gilbert (more on this later) who has inspired me greatly and who has shown the way forward in terms of making therapy a more client-centred, arguably more human experience than many have encountered in the past.

Stigma

It is well-known that people with mental health problems have long faced stigma within our society. This has, to an extent, been exacerbated by the popular media, both 'news' and artistic. Despite attempts to present mental health as being no different to physical health in the sense that it is a natural part of all of us and we are all vulnerable to ill health, as well as high-profile media personalities 'coming out' about their problems, to some extent, the stigma remains. The Equality Act (2010) sought to build on the earlier Disability Discrimination Act and as a part of this, ensure services treated physical and mental ill-health with parity. As even a casual observer can tell, this is not the experience of most service users. One of my reasons for writing this is to demonstrate the foolishness of such stigma. The people in the following case studies have no prior history of mental health problems. They are people from varied backgrounds and of varied ages. It is the essential 'normalness' of them that gives insight into what should be clear that mental ill-health can befall any one of us. For further reading on stigma, see Gibbons, B. Birks, M. (2016) Is it Time to Re-Visit Stigma? A Critical Review of Goffman 50 Years-On *British Journal of Mental Health Nursing: July/August 2016, Vol. 5 No. 4.*

Terminology

This book has been written with the mental health nurse in mind, therefore the use of terminology reflects this. That said, I have tried to keep technical terms to a minimum, in order to make the work as accessible as possible to the wider readership. One area which always raises comment is the term patient – should it be patient, with its rather paternalistic, medical model overtones, client, which is commonly used in therapy but has rather commercial connotations or service-user, which is preferred by many in today's recovery oriented environment. Without wishing to offend, I have used all three interchangeably, however, I am aware of the probability that I have likely caused offence to someone, at some point, somewhere. If I have, this was obviously not my intention and if you are still determined to be offended after reading this explanation, then I would respectfully suggest you seek suitably qualified help.

The following case studies have all been written using pseudonyms. Though the names are changed, all else remains as it happened.

Case Study One

Jane – Avoidant Personality

Jane was a 40-year-old woman, referred to the clinic by her GP with a rather exasperated-sounding letter. Jane had been complaining of persistent low-mood and anxiety for several years and had been treated with a variety of antidepressants to little or no effect. At that point, she was prescribed fluoxetine 20mg (aka Prozac – the lowest prescribed dose) but reported feeling "unsettled" for much of the time. On assessment, she presented as a pleasant, smiling woman, giving appropriate eye contact, in I suppose what would be considered smart-formal clothes. Jane related that she had felt low in mood and "nervous" for most of her adult life as far as she could recall.

Jane had no children and seemed frustrated when I asked if she was in a relationship with anyone. She readily admitted that this was a source of unhappiness for her, for she had never had a relationship lasting longer than a few months. Having gone through other assessment criteria for depression, her mood, concentration, drug and alcohol use (a glass of wine on a Saturday at home or on going out), risk of suicide or self-harm (none – she denied, I believed, there seemed no need to go further to formalise this), I asked her more about her current relationship.

The relationship with her current partner was described as "good – supportive". Jane went on to say "I hope I don't ruin it like I usually do". I asked her to say more about this, to which Jane replied that something she was in the habit of doing was texting her current boyfriend at work to make contact but that she would worry if she did not get a more or less immediate reply. This, it turned out, became a problem for many. Understandably, being contacted, often under the pretext of confirming plans for later, but sometimes just for outright reassurance, became quite irritating for many. However, perhaps surprisingly, it was always Jane who would terminate the relationship first. This, it seemed, was due to her inability to tolerate the uncertainty that she felt throughout the duration of the relationship. Relationships were what she craved but they produced huge amounts of stress for Jane, as she was constantly living in fear of imminent rejection. Therefore, she sought reassurance, which itself introduced strain into the relationships and fed Jane's fears, creating a vicious circle.

The more we talked the more it seemed this was a central part of Jane's current and indeed past psychological distress. Jane attributed this not, as may be assumed to a prior boyfriend, but to her mother whom she described as a "nervous, clingy woman".

One common misconception propagated by those relatively unfamiliar with CBT is that it ignores the past (Grant *et al.* 2008.). This is in fact not true – the past is acknowledged as important in shaping our present (Grant *et al.* 2008. BABCP, 2017), however, by itself, this knowledge is of limited utility. What is felt to be of more practical value than this insight alone is having skills to manage the challenges produced by this past and move forward with one's life (Kuyken *et al.* 2008).

As is recommended, a vital first step, especially with someone with little or no knowledge of CBT is to provide a background of what the therapy entails and how it works as a part of psychoeducation (Grant *et al.* 2008).

Taking a collaborative approach is vital in many forms of therapy and CBT is no exception. Jane was receptive to the observation that her anxiety was at least being exacerbated by the constant fear of rejection and the constant need for reassurance. We therefore discussed how long she might find it bearable to wait for a response post-text. In order to check this out, we decided to try a behavioural experiment (Grant *et al.* 2008). Behavioural experiments are commonly used in CBT to test a hypothesis which the client or service user my hold about their ability to cope and we will be looking at further examples of this in chapter. Behavioural experiments also help to establish a baseline, in order to then design a programme of graded exposure (Grant *et al.* 2006). The type of experiment we agreed was for Jane to text her partner and then switch off her phone to ascertain how long she could tolerate the uncertainty. As with all experiments of this kind, there is some degree of risk. Her partner may be too busy to respond or have his phone switched off. It may be that this is the last straw and he becomes angry at yet another text while he is supposed to be at work and respond negatively. The stress of the experiment might be too much for Jane and lead her to the conclusion that the therapy is too difficult. Actively avoiding uncertainty has become her coping style (Mittal & Griskavicious, 2014) and although on an intellectual level, she was aware of this, emotionally it is asking a great deal to expect change in the short-term at least.

Therefore, there are some very important 'rules' that need to be applied:

1) Always, always take your time to explain the rationale. It might seem obvious to you but to your client, it is often the first time they have encountered this approach. They will need a certain amount of time to understand the rationale behind it and may need more time to think about whether they want to do it. This is normal and ok. Expect it and do not under any circumstances be tempted to rush the process.

2) Make the experiment achievable. There is clearly no point in asking Jane to turn her phone off for five hours until her partner gets home. This would be flooding (psychological implosion) and although there is some evidence for its effectiveness in treating PTSD (see chapter), this is a technique which does not suit many clients and needs to be applied with some caution (Fisher & Sprich, 2016). This is an inexact science and people vary in motivation and distress tolerance. The best way is to be flexible and discuss it beforehand. Ask her how long was the longest time she had to wait to get a reply prior to this. Then ask her to rate how anxious she felt on a scale of 1-10 with 1= no distress and 10 = unbearable stress. Most people will choose a number in between and not the extremes, but if they do, that's ok.

3) With the rating and the time, ask Jane how she feels about perhaps adding one minute to the previous longest time. Does she think she can achieve this? This is as much a test of motivation as anything else. If she says no, go back to stage 1, perhaps adding a little motivational interviewing (Greenburger & Padesky, 2015).

4) What does she predict will happen? I can't overstate the importance of this. So often, in anxiety, we find that there is a tendency to predict the future (near & far) in negative ways. It is worth checking this out at initial assessment and whenever other opportunities arise, such as just before embarking on a behavioural experiment or exposure session. Keep a record of all such predictions (Grant *et al.* 2008) as this will then provide evidence to the client of how predictions play a role in maintaining anxiety and how we often predict consequences which are many times more negative than they actually turn out to be. This in turn leads to avoidance and reduced opportunity for disconfirmation (Moghaddam & Dawson, 2016). Better still, check with the client if they experience any imagery to accompany their predictions – for example, her partner packing his bags and leaving or with another woman (something Jane often expressed a fear of).

5) Check medications – prescribed or otherwise (including alcohol). There is no point in encouraging exposure to anxiety if the anxiety is moderated by such means – it defeats the object. Purist therapists will often refuse to assess clients while they are taking anxiolytic medications.

6) Usually it is best not to try any of this this at all until you are confident there is at least a developing therapeutic relationship. You are asking a lot of your client and they need to trust you. It might go wrong and you may need to lower your aims at least temporary. This is only going to work if there is at least some trust. Otherwise you risk losing the confidence of your client and potentially putting them off therapy. Pick your time carefully and don't be in a rush to 'prove a point'. This is about helping your client, not showing them how clever you are.

7) For this experiment to work, Jane's partner needed to be kept in the dark – this wasn't going to be convincing if he knew how to react, though there will be times when a certain amount of collusion is required and necessary.

Having agreed to try the experiment, Jane texted in the usual manner and then switched her phone off. We then waited – it was vital not to distract her at this point as that would be avoiding – however, in some cases, this may be necessary – even desired. Some therapists feel that avoidance, as a safety behaviour, is always problematic, however, others will take a more flexible, possibly pragmatic stance (Grant *et al.* 2008).

My feeling is that this is one of those situations where your experience and your client will guide you. If you have done a good enough job of educating them and they are motivated, they will make their preferences known. If it is a choice of distraction or no therapy, then choose distraction as a means to achieving a more purist approach further down the line. If it helps with initial engagement, it can be a helpful starting point – but – it requires honesty and full collaboration from both parties as eventually, the non-distraction option is the one which provides the best prognosis (Barker. 2008).

Jane managed a full five minutes the first time. Over the next three weeks she managed ten and then fifteen minutes. This was something she clearly found very difficult but she was able to see that initially, she had predicted she couldn't cope for more than two minutes (though in reality, the longest she had waited was five minutes before getting a reply pre-therapy).

After six weeks, Jane was reporting a significant reduction in her levels of anxiety both generally and specifically around reassurance-seeking which she had almost completely stopped doing – at least in this form. Her mood had lifted and she felt physically less tired and stated her relationship with her partner was improving. This was the point at which Jane needed to decide whether to continue with therapy and explore the wider origins and manifestations of her anxiety or accept that she had already achieved more than she had hoped for and for now at least, disengage.

She chose the latter and to this day, with her increased knowledge of herself and of the nature of anxiety, manages to live a full and (mostly) happy life.

There will be some who find this frustrating – that she didn't achieve a moment of clarity – a "corrective emotional experience" (Weatherhead & Flaherty-Jones. 2012) in psychodynamic terms and put to rest the source of her 'neurosis'.

However, if we are sincere as practitioners in our desire to provide client-centred interventions and patient choice, we have to come to terms with the fact that we are not the ones who decide what is best, we merely facilitate informed choice.

Phil – Panic Disorder

My first encounter with Phil was via a phone call from his mum at 9.30 on Monday morning. This was significant as it transpired that Phil had been refusing to go to school after several weeks of reporting feeling very unhappy at school. It turned out that his mum, Carole was quite familiar with CBT having been successfully treated for anxiety herself some ten years earlier. She had taken Phil to the GP but as is often the case, was disappointed that all they could offer was medication and/or a five month wait. Clearly, at any age, five months out of school is a very significant loss of learning, but Phil was 15 and about to enter what is often portrayed as the most vital part of his school life so in many ways his anxiety was shared by his parents hence the call to a private therapist.

I agreed to meet the family at home for an initial assessment. Not being used to working with school children I was apprehensive but Phil's mum was so in need of some intervention on his behalf I agreed to see him.

Arriving at the house I was greeted by Carole and ushered in. Carole was joined by her husband and Phil and although there are times when it is far preferable to be alone with the client, in this situation, I felt for now, having his parents there was right – for now.

In situations where clients are joined by well-meaning parents/partners/friends, the voice of the client can sometimes be lost (Tambuyzer *et al.* 2011).

However, there is no hard and fast rule to this (Tambuyzer *et al.* 2011) and certainly in situations where there are children or a male nurse with a female client, it is sometimes prudent to have a chaperone so long as they do not impede or inhibit the session.

Phil's mum had already given me quite an extensive background during the phone call, however, I asked for details again now that Phil was in front of me. Phil had been experiencing somatic feelings of anxiety (churning stomach, dizziness, sweating, dry mouth, nausea) when at school. He also experienced what Carole described as panic attacks in one class in particular, the science class. This had happened "2 or 3 times" and now Phil was experiencing feelings of panic and dread on the way to school and had recently refused to leave the car when they arrived. His dad had responded angrily at first and basically shouted at Phil to get him to go into class but this resulted in Phil phoning his mum at her workplace in "extreme distress" pleading to be taken home. This meant Carole had to leave her workplace and pick him up, take him home. Crucially, the following Sunday evening, Phil had a panic attack at home – which often happens and goes some way to proving to the client that the location is not the trigger they thought it was.

During the assessment, Carole did most of the talking. I did try to check much of the detail with Phil including asking for details only he could verify, such as where, when, how often etc. Apparently the school were aware and the teacher in the science class had "also been treated for anxiety".

I asked Phil to describe what his panic attacks were like. He described the list of symptoms his mum had related to me previously.

The situation appeared to be:

Phil feels anxious about going to school

Phil has experienced panic attacks at school and at home

Phil is now missing school time

This is worrying for his parents and is disrupting his mum's work

In a clear and understandable attempt to avoid the unpleasant sensations associated with panic attacks, Phil was afraid to go into school. This is due to association and misattribution. What was important was to demonstrate to Phil that his panic, whilst *seeming* to be associated with the science class because that is where he first experienced it, was in fact independent of that. Experienced therapists know that sooner or later, the client will experience panic symptoms in other places (Grant *et al.* 2008). It will be a temptation to avoid these places too until ultimately, the client retreats into their own home, fearful of going out at all, gaining a diagnosis of agoraphobia.

Of course, in the end, the unfortunate client experiences a panic attack at home and realises too late that it wasn't all about the location at all. But by then, the damage is done, they are often fearful and severely lacking in confidence. This is a horribly disabling condition and early intervention, as in so many things, is vital (Wills, 2008).

As usual, after a full assessment, it was appropriate to inform Phil of the way CBT worked and how it could be used to help him with what we agreed was a diagnosis of panic disorder. Carole was an advocate here as she had previous positive experience of CBT and this made the whole process somewhat easier than may have been the case otherwise.

It is vital that clients are able to give informed consent, not only to assess motivation but also in order to achieve the best outcomes (Fisher & Oramsky, 2008).

After a discussion of the nature of anxiety – physiological aspects, complete with handouts and drawn diagrams, Phil disclosed that he was frightened that he might stop breathing. This was due to the tightening of the throat experienced by some people experiencing acute anxiety, along with difficulty in swallowing – which was making him reluctant to eat. Phil reported that this was only an issue at school – at home, he did not experience these problems.

Clearly then, the problem was centred around the school.

There were, as we saw it, 3 options:

1) Phil changes school – this was not a popular choice from a practical perspective as the nearest alternative was several miles further away. From a CBT perspective, this would also be undesirable because it would reinforce avoidance (Wells, 2008). Phil would also miss his friends and social life. He was a promising footballer with the school team.

2) Phil is taught at home – again, impractical as both Phil's parents worked and again, there were the problems of avoidance and loss of friendships as above.

3) Phil learns to cope with his anxiety and goes back to school – in many ways the most challenging option for Phil, but realistically, the only solution, especially with GCSEs fast approaching. We all agreed that if the problem could be 'nipped in the bud', this would be the least -worst option. Even Phil agreed to this, so we set about formulating a plan.

CBT often uses a 5-areas formulation to illustrate clearly to the client and therapist what the current situation seems to be. (Greenburger & Padesky, 2015)

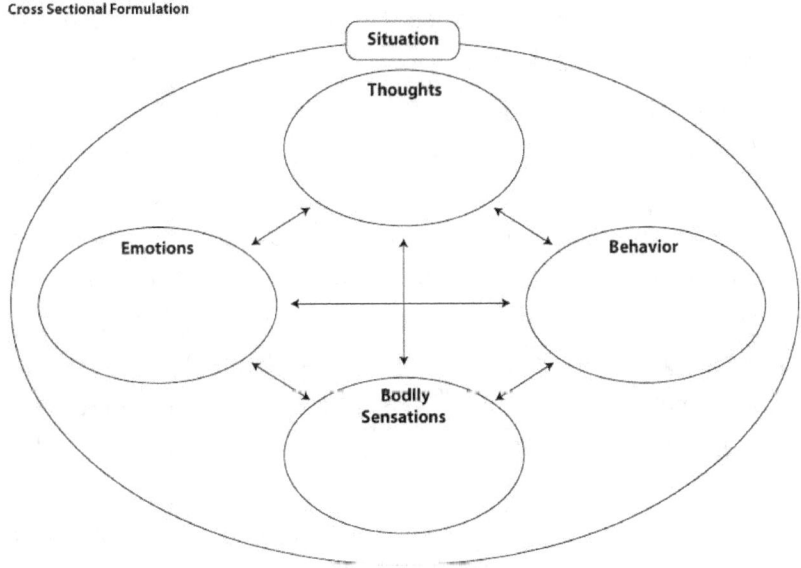

Cross Sectional Formulation

Fig. 1. Five-areas formulation.
Greenburger & Padesky 1995

As we can see, Phil's situation at the current time was becoming quite stressful even before he became anxious about becoming anxious. It is important for the client to understand this formulation as it places context onto their current difficulties, which for many clients can appear to come "out of the blue" as it were, in turn, encouraging catastrophizing and focusing on physical sensations (Wills, 2008).

A common trait in most of us when experiencing anxiety is to focus on physical symptoms and sensations, so well documented in anxiety (Wills, 2008). When we do this, of course, they tend to increase, thereby making us focus on them more and so on. We convince ourselves that something terrible is happening and self-diagnose. By now, it is very difficult to think rationally and we start to panic. Of course, as distressing as this is, it can only last for so long as we are physiologically incapable of maintaining this level of adrenaline indefinitely, as the diagram below shows.

This knowledge is a vital part of the beginning of therapy and must not be rushed. Psychoeducation has been shown to improve outcomes and is a really effective way of establishing rapport and credibility with the client (Wills, 2008).

Now that the formulation is ready and discussed and agreed with, the next task is to agree a plan of how to progress. The agreed goal was for Phil to return to school as soon as possible. We were very fortunate in having motivated help not just from Phil's parents but from the school and individual teachers. Without this, the next step would have been much more complicated. Paul Salkovskis states that it is one thing for the client to understand intellectually but to see true change, they must understand emotionally too (Wells *et al.* 1995). In panic disorder, the client will focus on the environment where they happen to find themselves when they experience panic and misattribute this to be the cause. In fact, panic disorder is largely maintained by fear of panic attacks (Westbrook *et al.* 2007) therefore exposure to panic attacks tends to reduce their potency to cause fear and thus the number of panic attacks the sufferer experiences (Westbrook *et al.* 2007).

I discussed with Phil simulating a panic attack. He agreed to try so Phil, Carole and I all engaged in over-breathing in order to demonstrate how the symptoms of panic are induced by this very simple exercise. Kuyken et al. demonstrates this very effectively (Kuyken *et al.* 2008) all the while communicating with the client, asking for feedback and in 2 or 3 minutes, the client is very light-headed and dizzy. Now of course this feels somewhat different when done in a controlled way but the message is clear – I felt the same symptoms, I didn't die.

Some clients will not be convinced by this due to its artificial nature, so before the exercise, we explored Phil's thoughts around the initial panic attack. I asked him to recall any thoughts and feelings he was having and write them down. I also asked his to rate on a 5 level how much he felt he was going to die, what he thought was happening and how much he believed this etc.

We repeated this exercise after the induced panic session and the scores, as they so often are, were lower.

I then asked Phil how he accounted for this – he gave the answer I had been hoping for – "because I now know it was only panic". Music to the therapist's ears, as real progress and understanding are taking place. So the next stage was to try to return to school armed with this new insight.

I should add that alongside this, we worked on some mindfulness techniques as an extra tool in the chest, as it were. These were for the night before and in the morning on the way to school. If necessary, they could also be used in the classroom – though there was some acknowledged risk in sitting in a classroom appearing to be daydreaming so we rejected that part of the plan.

Phil's teacher had, through kindness, offered to allow Phil to leave the room if necessary, minimising any fuss or attention that might make him feel worse. While this was done with the very best intentions, I discouraged this as an option because it was, in the end, simple avoidance and would hinder recovery. Phil was keen to keep this option, however as a safety-seeking behaviour. Therefore, I had to point out that although there were valid reasons for staying put and in my opinion, he should do so, if it got him into the classroom, we could keep that option for now.

Three months on there have been some hiccoughs. Phil was still phoning his mum at work, another safety-seeking behaviour so I had to ask that we ban this. Having agreed, he then proceeded to phone his dad at work. He was not asking to be brought home, merely seeking reassurance, which I suppose at age 15, should not be surprising. I wanted Phil not to do this though. Especially as doing so clearly irritated his dad who would then shout at him. He could manage, he just needed to believe he could manage.

Phil wrote some cue cards to read to himself – in his own handwriting, with reassuring reminders such as "It is only anxiety" and "This will soon pass".

His road has not been smooth – not many are – but he has coped admirably and shown considerable maturity beyond his years. His parents asked to discontinue therapy at this point, citing the cost as something they could not maintain. I keep my costs as low as I can – much lower than many and even travel to people's own homes to work with them. They seemed a reasonably affluent family living in quite an expensive area but I respected their situation as they were trying to get Phil onto a waiting list at the local GP and I was always going to be a stop-gap until they got an appointment. At the time of writing, a 5-6 month wait seems to be typically what I hear from those seeking help. At least Phil should be in a good position to progress from here and with the help of Carole in particular, the prognosis is a qualified fair-good.

Moira – Social Phobia

Moira came along with what, to many people, may seem like a fairly unexceptional, even trivial, problem. Little by little, over the last few years, she had become more withdrawn and isolated. She explained that her husband had died three years previously and her children had all moved away because of work and/or marriage. This meant she was alone at home but she had friends around her, at least initially.

Moira was a woman in her late 50s who had worked hard all her life as a cleaner with the local council. She had given the job up around six years ago to care for her then ill husband. She had done this without question – she was as she put it, "not getting any younger" and the job was physically demanding. Her husband had been a miner but had been made redundant in his 40s and had never found work since, despite his best efforts.

Having developed COPD, (chronic obstructive pulmonary disease), no doubt as a result of his job but also a heavy smoking habit, Moira saw no alternative but to care for him full-time.

Now that her husband had gone, Moira decided she needed to get on with life. However, she had recently been conscious of finding it difficult to go out. In particular, she found women's clothes shops a real challenge. Moira stated that she never felt herself to be particularly self-conscious but she found herself one day in a shop asking an assistant ("Just a child, really") if she could try on a jacket. Moira felt herself "colouring up" for no explainable reason.

Thus, Moira's circle of friends and acquaintances had slowly diminished. She did have some friends, it is true, but as often happens after a bereavement, many of these had fallen away since the funeral and so she was living most of her life, essentially alone. Everyday jobs had become more of an effort and there was evidence of a mild depression developing. One thing that Moira had not changed though was her standard of appearance. She always took the time and trouble to present herself immaculately and was casually but smartly dressed as we met.

Then there was the incident in the clothes shop. Moira described how it was just a normal day and she decided to do some shopping. She decided she didn't want to dress "like an old woman" any more so she went into an attractive-looking shop she had never been in before. She said she "felt the stare of the young assistants" on her as she browsed. She felt that she was being appraised unfavourably. That the staff were thinking "who does that old woman think she's kidding"? Objectively, Moira could now tell herself this was nonsense, that it was her own lack of confidence that was saying those words to her.

Much of the work we would do would be around self-esteem and a tendency we all have, whether depressed or not, to project our anxieties onto others (Weatherhead & Flaherty-Jones, 2012) and to mind-read when we are feeling vulnerable (Gilbert, 2007).

We agreed that the way forward was that a full physical examination should be conducted by Moira's GP. Many symptoms which mimic depression and anxiety can be accounted for by underlying physical causes and Moira would not have been the first woman of her age to suffer from hot flushes.

It was in any case important to explore some of Moira's worries so that she could start to integrate again. Since this incident, around six weeks prior, she had barely left the house and had not spoken with anyone except her GP – the source of the referral.

We both agreed on the course of treatment, cognitive restructuring (Leahy & Holland, 2000) to develop skills to challenge automatic thoughts and assumptions and ultimately, core beliefs (Wells *et al.* 1995) and a behavioural approach – graded exposure (Wells *et al.* 1995). A recurring theme in our discussions was an assumption made by Moira that others were judging her. This is common in people in whom self-esteem is low, specifically in people with a physical impairment, as we age or in people who are more socially isolated (Norton *et al.* 2005). Despite some subjective experience to the contrary, it is common to over-exaggerate the extent to which others place importance on how we look – unless we do look extraordinarily unusual – and Moira didn't. She was very pre-occupied with the likelihood that she would blush. Using downward arrow technique (Greenburger & Padesky, 2015. fig 2) it is possible to use Socratic dialogue to enable toe client to discover a more rational response. A behavioural technique known as shame attack (Phillips-Sheesley *et al.* 2016) is a more dramatic and indeed, dramatic-sounding accompaniment to this. This is not to be confused with the ground-breaking work of Paul Gilbert on shame and guilt (Gilbert, 2000) – this is completely different. If you haven't come across Gilbert's work, I strongly recommend it. There is a lot out there – there are even video lectures on YouTube which are very easy to follow and really resonate with service users and therapists alike.

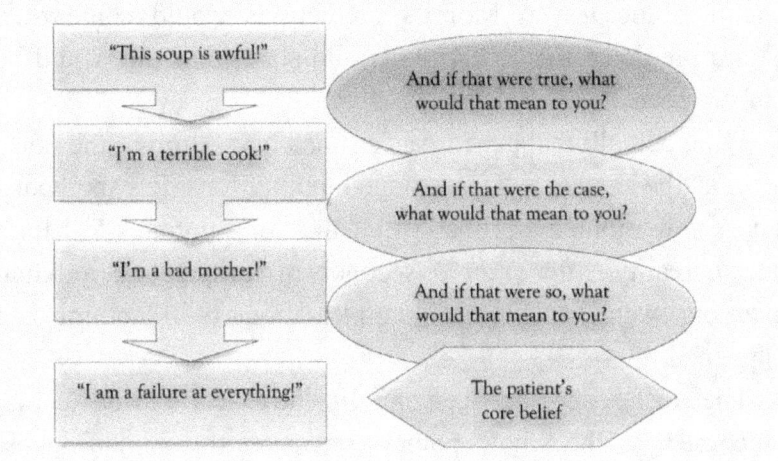

Fig. 2. Greenburger & Padesky 1995

Shame Attack is, let's face it, not an attractive-sounding intervention. Behavioural Experiment is arguably less aggressive-sounding and is perhaps a more scientific description. For this experiment, we chose the university canteen at lunchtime. Possibly the noisiest place around and with around 200 or so students all chatting, eating and playing with their phones.

I purchased, with Moira, two cups of tea. We had previously gone through the plan as to not do so would have been non-collaborative, unfair, and counter-productive. With the two cups of tea on saucers on the tray, we walked towards the rowdy and mostly occupied tables. When there was a bit of space and the likelihood of avoiding spillage damage, I dropped the tray – crash!

Beforehand, we had discussed the rationale for this. To draw attention and test if Moira's predictions would come true. Moira predicted that everyone would stare, that she would be embarrassed and feel shame.

What actually happened was a momentary drop in the noise level as those nearest to us stopped talking for a few moments. Those few moments probably felt like long minutes to Moira but in reality, after 4 or 5 seconds of turning and looking, everyone went back to their conversations as if nothing had happened.

The difference between predicted outcome and actual outcome was the whole point – it wasn't the unmanageable disaster Moira had predicted and it provided a basis for not just immediate de-brief but further work on negative thoughts, predicting the future and making assumptions.

This provided not just an opportunity to gently challenge some of Moira's pre-existing assumptions about what might happen and how she would cope in an embarrassing situation, but it represented evidence that her fears were largely unfounded. The risk here is that when reminded of how she coped, Moira could simply say "ah yes, but you were there" to which I could reply "but what difference did I make"?

Moira might also say "it was ok this time but what about next time"? Of course, there are no guarantees in life and this should be emphasised. For many people, anxiety comes from lack of certainty. As difficult as this is, we are all in the same situation and giving reassurance for something we cannot control is foolish, however well-meant. A therapist or nurse can no more predict the future than their client, however, they can gently encourage the client to draw on their previous experience. The intervention can go something like;

"We cannot know for sure what the future holds, but how did you cope last time"? "What did you do to manage last time this happened"? "How did that turn out"?

The fact that they survived shows a level of coping the client might not have credited themselves for. Of course, some will dismiss this as mere luck. Even here though, the likelihood is that there were other factors involved and/or they have coped on several occasions, making the luck explanation slightly harder to justify. It is important not to get drawn into an argument here. If the client is holding onto their anxiety to such a degree, all that is needed are more opportunities to disconfirm, which they will either agree to or not.

At times, this level of 1:1 work with clients can feel like a game of chess. The nurse or therapist makes a suggestion or observation, the client, counters it – often with a "what if"? This is where much of the skill lies. It can be very easy to get drawn into a frustrating, adversarial situation of 'punch and counter-punch' but this is completely pointless. All that will happen is the therapist/nurse gets frustrated and the client feels browbeaten or bullied.

As the therapeutic relationship developed, it became easier to anticipate Moira's responses and as with all good relationships, it became easier to counter the what ifs with a smile or a nod so she became more adept at countering herself – I didn't have to. An effective therapeutic relationship such as this isn't created overnight, but it does make the process far more effective, efficient and less stressful for both. The mutual respect of a relationship such as this makes for far better outcomes (Barker, 2008). It won't exist in every encounter with every client but it is worth striving for and is another example of how so-called 'basic' skills are essential as a foundation for the use of more advanced interventions.

Chloe – agoraphobia

There can be few things in life more frustrating to everyone involved than an inappropriate referral. It wastes everyone's time and damages trust. Chloe's referral was one such example. Almost weekly, we would receive a referral from primary care for someone with agoraphobia. We could guarantee they wouldn't attend. When someone has found it impossible to leave their home for several years in some cases, clearly, they are not going to be able to attend an appointment at a busy inner-city community mental health unit – it's just not sensible to ask them to.

Yet on a pretty much weekly basis, this is what happened. We would schedule the assessment, knowing full well the chances of non-attendance would be very high. Then we would write to the referrer to explain they had not attended. Our time is wasted, the referrer's time is wasted, they are frustrated and the patient is left feeling poorly understood, probably embarrassed and frustrated too, their already fragile self-esteem further decreased.

We would advise referrers of this when writing back to them but still the letters came. Therefore, we tried a different tactic – one that many services probably couldn't offer today – we offered to visit them at home. It seemed such an obvious next step and Chloe was the first.

When I went out to see Chloe, she only lived about a mile from the unit, though it might as well have been a hundred miles as far as she was concerned. Chloe hadn't left the house for two years except to visit her GP on the next street some weeks earlier. She had done this with great reluctance and with much support from her partner as they wanted to get married in three months- time but could see no way of achieving this at present.

Chloe had found life a real struggle since being experiencing a panic attack while out shopping about two and a half years ago. At first, she had tried to forget about it but then the classic cycle had kicked in. From the initial panic attack – which like most people, she had misattributed to a physical health problem and/or the setting, she had started to avoid crowded places and her circle of activity had reduced. She had given up her part-time job and relied heavily (financially and every other way) on her partner, Steve. Understandably this had put a strain on the relationship but they were a very committed couple and had recently decided they wanted to marry. Both being in their early 30s, they had come to the conclusion that life was happening around them. Their friends were all marrying, having families and travelling and they felt they were missing out. Chloe also felt that having a target to aim for would provide her with added motivation to overcome her fears.

There was much to admire in this young couple's relationship and their devotion to each other. This meant therapy would have a very good chance at success. The only concern I had was the wedding had been planned for three months' time – a very short period in which to achieve her goals. The invitations had been sent out, deposits paid and the hall booked. This level of commitment was admirable but was it realistic?

CBT emphasises the need to work collaboratively with the client, in order to empower them to achieve their own goals, not the goals of the therapist (Grant et al. 2008).

It is important, when constructing a hierarchy of goals, to be guided by the clients aims. They have to be relevant to the client to have meaning (Grant, et al. 2008). This is not to say that the therapist or nurse cannot guide – they can. It is important, for example, that the relationship be an honest one – for example, I expressed that I was concerned that 3 months was going to be a very tight schedule when Chloe had not (until the GP visit) been able to leave the house for two years. I was concerned that she was setting herself up for disappointment. In the end, the humbling truth was, Chloe managed to achieve her goals due in the greater part to her motivation and courage. We started in the usual way by collaborating to produce a hierarchy of exposure. This needs to be done with care and the option to revise must be available. The stages of exposure need to be difficult enough to be of benefit but not so difficult that they seem unachievable. For Chloe, stage one was to walk to the end of the front garden path (about 30 feet) and rate her anxiety. If it was above 4, she would stay until it came down to 4. This follows the usual principles of exposure as stated by Leahey & Holland (2000) and no distractions or safety behaviours are allowed. These days there are many ways in which someone might be able to avoid their anxiety by use, for example, an i-pod. While this may at time be useful and even advised, in circumstances such as these, it would prove counter-productive.

After consistent reporting that the end of the path was no longer producing higher anxiety ratings than 3 or 4, the next step was the corner shop around 100 yards away. Then the bus stop another 100 yards away and so on. Each time, practicing twice per day minimum, staying until her anxiety levels came down to 4.

These activities were chosen because they appeared to be 'normal' activities. Often, clients can feel very self-conscious and do not want to draw attention to themselves as this tends to increase their worry and anxiety. It was important therefore to structure the exposure sessions to appear innocuous but be at a time of day when such activity would not seem odd and be worthwhile from an exposure viewpoint, so 10am and 4pm were chosen. This avoided the worst of the rush hour but was sufficiently busy to not draw attention but be therapeutically purposeful (Clark *et al.* 1999). It is important that the therapist is open to compromise at times like this. It would be easy to go straight for the busy periods and create almost a flooding-type scenario. However this would be to risk the therapeutic relationship and set the client up for probable disappointment. The risks and benefits need to be discussed and weighed carefully. Mistakes can and will be made and this is why it is important to be open and honest with the client. People are individuals and this is, to a degree, a rather inexact process. Even when the 'right' balance is achieved, there will be good days and bad days and these need to be anticipated so they don't come as a surprise.

However, as I mentioned, Chloe was exceptionally motivated. My input was mainly to educate, discuss, encourage and support. Chloe did all the hard work and the wedding day was a success for all. A relatively easy (for me) process turned into one of the most personally and professionally satisfying cases.

Lauren – OCD.
The burden of the nurse

Lauren came to therapy as a result of attending a class I was teaching on graded exposure. It has been a feature of teaching such classes over the years that almost every time I teach, at the end, someone either hangs back or emails me afterwards to ask if I "do private work".

Lauren was a year 3 student mental health nurse whom I knew but only slightly. She fell into the 'hang back at the end of the class' group at this point and when everyone was gone, said she suspected she had OCD based on how I had described it during the class.

OCD is characterised by:

> "recurrent obsessional thoughts or compulsive acts. Obsessional thoughts are ideas, images or impulses that enter the patient's mind again and again in a stereotyped form. They are almost invariably distressing and the patient often tries unsuccessfully to resist them. They are however, recognized as his or her own thoughts even though they are involuntary and often repugnant. Compulsive acts or rituals are stereotyped behaviours that are repeated again and again. They are not inherently enjoyable, nor do they result in the completion of inherently useful

tasks. Their function is to prevent some objectively unlikely event, often involving harm to or caused by the patient, which he or she fears might not otherwise occur. Usually this behaviour is recognized as pointless or ineffectual and repeated attempts are made to resist, Anxiety is almost invariably present." (Bream *et al.* 2017).

For Lauren, this meant many of her job-related tasks were taking a long time and she was filled with doubts over her ability to perform to the expectations of those around her as well as herself. This was also having what she thought was a negative impact on her home life. When I asked for some examples of this, she cited a recent situation where she was sending some coupons off in the mail (this was 2005) in order to receive a special offer. This was to obtain a birthday present and she was anxious that this should be done correctly. She had checked the coupons, the address and the post times. Then she placed the coupons and completed form in the envelope with the address on and placed it on the shelf for posting tomorrow. Lauren described how that evening, she felt compelled to open and check the envelope 8-10 times. She knew, logically that this would be unnecessary yet she couldn't relax until she had checked and this relief only lasted for 20-30 minutes. Of course, this wasn't just frustrating for Lauren but her partner and the knowledge of this made her feel even more anxious, which increased the checking and so on.

When deciding where to start treating a client, it is helpful to elicit their priorities (Grant *et al.* 2008). Commonly, clients will state, when asked, their ultimate goal, which could be (and often is) as nebulous as "I never want to feel anxious again". While this is completely understandable, it is also completely unrealistic. There then follows a process of negotiation between client and therapist. This often includes a measure pf psychoeducation (Thomas & Drake, 2012) – this is a tactful way of the therapist stating "it doesn't really work like that" and re-centring the client.

As discussed previously, at times, a hierarchy of exposure is fairly easy to construct – collaboratively, of course. At other times, the process is less obvious. In Lauren's case, although her checking behaviours were not confined to the contents of envelopes, this seemed as good a place as any to start. During our initial assessment, Lauren had told me that on her clinical placement, some of the doctors would leave doors and windows open and/or unlocked as they went about their duties and if she left work without ensuring they were all closed and secure, she would be asked to account for it by the unit manager. She therefore made it a priority to ensure everything was secure – no matter what else she had to do at that time. Also, in common with all nurses, she was acutely aware that any drug administration error – however harmless or innocuous – would result in an investigation, copious paperwork and the obvious disapproval of her manager. In reality, those of us with such clinical experience know that all nurses at some point in their careers will make such an error at least once and often many more times – such is the pressure they are under and the demands of the role. 99 times out of 100 it is so minor as to tempt the nurse to remain silent as no-one, not even the patient would probably notice and no harm would result.

However, it is incumbent on nurses to self-report such errors. Furthermore, it is common practice in many trusts these days to require the nurse to make a personal and written apology to the patient – even when the patient knows nothing of the 5ml extra lactulose they had in the dispenser cup that morning.

No surprise then that the stress of having to get it right, all the time and every time can take its toll. This is a serious business that could have serious consequences and the thought of making mistakes weighs heavy – especially when there is pressure to administer medications as quickly as possible – mindful of the need to 'muck in' with the rest of the day's duties within an often busy team.

There are ethical implications to asking a nurse to deliberately commit a drug error which made this approach something of a non-starter. Therefore, we decided to start with the envelope situation.

Although whenever possible, it is preferable to expose the client directly to the situations in which stress occurs, this is not always necessary (Bream *et al.* 2017) or possible.

Sometimes, the techniques learned in learning to cope within a given scenario can be applied to others – generalised outwards as it were.

Therefore, we decided to use the envelope scenario. There is a temptation is such situations to manufacture a reason to send something to someone just to re-create the physical setting. This stands lees of a chance of success because there is no' jeopardy' - no anxiety is created is, for example, I ask Lauren to send something to herself – if it goes wrong, who cares?

So, it had to be real. Lauren had to send a payment by cheque for some item she had bought so we decided this would be the perfect 'excuse'. I decided I would not watch her do this as she would be able to ask me for reassurance and this would defeat the object of the exercise so it had to be done with me out of the room. Having written the cheque and sealed it in the envelope, I asked Lauren not to post it until tomorrow. Being a student mental health nurse, she was familiar with relaxation techniques and mindfulness practice. If not, we would have had to work on coping mechanisms such as these and some others first.

Although Lauren found the task predictably difficult, her preparation and determination saw her through. The process of exposure to the anxiety taught her how to cope. She would use distraction – though some therapists view distraction as avoidance. My experience is that in the real world, this is a useful tool if used sparingly. Clearly, at times, distraction is much more of a challenge, such as when she went to bed. Again, preparation is key. We had foreseen this and her preparation was to write down what she could do if it turned out that there was a problem with the cheque or it didn't arrive. I asked her what might happen under such circumstances. Her response was she thought she would get a letter or phone call informing her of this and she could apologise and send another. She had not articulated this out loud before and felt her anxiety was around vague, unclear thoughts around defaulting on the payment and anticipated anger on the part of those she had to pay. When I asked her how realistic this was, she conceded that it was more likely that a professional company would be more tolerant of this than an individual as it must have happened to them before. This enabled us to reason that it would be annoying to all concerned but not disastrous. This enabled Lauren to cope with her anxiety as she felt the consequences were – in a realistic appraisal – not as catastrophic as she had imagined. Catastrophising is something we all do (Grant, *et al.* 2008) and often leads to unrealistic assumptions which can envelop the client in further anxiety. A realistic appraisal can help to separate those things which need our immediate and full concern from those which do not. Therefore, the long list of things which can make us worry becomes a little more manageable. This skill is a vital by-product of the process of graded exposure and one which the client can then use in future.

Clive – PTSD

At the start of my career I had no great ambitions to work as a therapist. I qualified in CBT because I was attracted to its relative simplicity, its pragmatic – even common-sense approach and most of all, I had seen it work and work very well. Occasionally, overstretched colleagues would ask if I would see someone they couldn't. This was how I came to see Clive. Clive was a GP who, around 25 years earlier, had served as a medic aboard the hospital ship Uganda used in the Falklands war. He had been experiencing what seemed to be classic signs of PTSD but was not certain as this was so very long ago. I met Clive at his surgery after the last patients had been seen. He described to me how recent military activity at the army base near to his home had triggered flashbacks and feelings of intense anxiety. Clive lived near a base which flew the large, twin-rotor Chinook helicopters and recently they had been flying quite low over the village. He described the first time this had happened several weeks ago. Clive described feelings of a classic escape/avoid appearance. He was with his young daughter and had felt the overwhelming urge to hide, so they ran into a pub and he hid in the corner until the feelings had subsided. Feeling understandable rather foolish, he shrugged it off as a sign of stress and got on with his life – until it happened again the following week and twice the week after. Each time, he had experienced feelings of intense panic (raised autonomic arousal), lots of physical sensations such as palpitations, shallow, rapid breathing and sweats. Clive denied seeing or hearing anything that wasn't there but he had already made the connection himself at this point as each time, it was triggered by the low-flying helicopters. Live described how as a young army doctor, he had been sent to the Falklands to treat the anticipated wounded. Ostensibly outside the immediate danger area, miles out in open sea, the hospital ship seemed safe. Receiving badly wounded and disfigured men at a moment's notice didn't faze him – "it's what I was trained to do" he said. He first noticed a change in his feelings after the sinking of HMS Sheffield and the Atlantic Conveyor by Exocet missiles. The notion that a jet, many miles away could fire a low-level missile that flew beneath the radar and literally blew ships apart with no warning was terrifying. Clive described his cabin, which had a porthole right in the middle of the red cross painted on the side. Being a hospital ship afforded no guarantee of protection – to the jet pilot launching the missile from 20 miles away, it would be just another ship to be attacked. Nothing to see, nothing to hear until Bang! To live from day to day with the constant fear of death at any second is something most of us will thankfully never experience. After weeks or months of this, it is hardly surprising that this kind of pressure takes its toll. That it should manifest in PTSD is hardly surprising either – the 25-year gap is unusual but not unheard of (Kuyken et al. 2008) – especially when the trigger is something specific and relatively uncommon (Kuyken et al. 2008). Helicopters were in short supply in this campaign. Those who remember this war will recall the famous 'yomping' across the main island by British soldiers who had to carry their kit and fight their way over to Port Stanley because most of the heavy helicopters were lost with the sinking of the Atlantic Conveyor – all the eggs in one basket indeed! Those helicopters that were available were used to ferry the wounded out to the hospital ship.

PTSD or PTSS (Kuyken *et al.* 2008) as it is known in America has received much more attention sadly as a result of armed conflict. It would be a mistake, however, to think it is solely a by-product of such violence. It is quite common for people to develop the disorder after a car accident, burglary, mugging or even just being a witness to these things. The 'common-sense' attitude to witnessing such events is to try not to think about it. This seems eminently sensible, however, unfortunately, this can be futile as the thoughts will intrude whether one consciously thinks about them or not. Ask anyone who has ever been on a diet how easy it is not to think about food – a relatively trivial example I know, but illustrative of the futility of trying to block out thoughts one is trying to avoid. In WW1, long before this process was understood, it was known in the early 20th century as shell-shock. Horrendously, troops displaying such symptoms were sometimes branded as cowards and even shot for desertion. Now we know that there is psychological damage occurring – a failure of the brain to process what it has experienced and almost a loop played in the mind, at seemingly random intervals. Obviously, even if one is aware of this process, this is still a very frightening and distressing experience. At its worst, there are accounts of the sufferer completely losing any sense of reality and lashing out causing injury and/or death to themselves and/or others. Not surprisingly, there is strong evidence to show such sufferers often self-medicate with drugs and alcohol – a maladaptive coping strategy particularly prevalent in former service personnel (Taylor, 2017). Again, this is hardly surprising when one considers the mindset required to function effectively in highly dangerous, stressful environments would involve conscious blocking of emotions – perhaps later relieved by letting off steam in the bar with comrades. This is one of the challenges faced by therapists. Many sufferers don't seek help and choose to avoid – perhaps feeling that bringing all the trauma up would serve no purpose. The stiff-upper lip, so useful in times of crisis, so valued by some as a sign of 'Britishness' can be a massive obstacle to receiving more long-lasting peace of mind.

Thankfully, today there is much more help provided by military and voluntary services. We still have a long way to go but the picture is vastly different to WW1 and there is generally much more readiness to seek and accept help. Clive was never going to be as challenging to treat as some ex-military personnel I had encountered, who found my non-military background a credibility-problem and used this to reinforce their reluctance to engage. Clive, as a medical professional knew about CBT and wanted to use this opportunity. CBT is not the only treatment found to be effective in cases of PTSD, EMDR (eye movement desensitization and reprocessing) is also found to be useful (nice.org 2017). However, Clive understood well the 'mechanics' of the treatment. Re-living is a treatment with well-proven efficacy. I had previously used the treatment with survivors of sexual abuse who experienced flashbacks and other psychological symptoms and had seen how challenging it was for the client to engage fully.

Re-living is a kind of graded exposure in vivo (Bowyer *et al.* 2014). It involves the client talking through – describing what they are going through not as a matter of historical recall but as though it were actually happening, there and then. It has to be done this way to be effective (Bowyer *et al.* 2014). Clients will often avoid doing this at first – either because it hasn't been explained to them properly or they find it too difficult. They will commonly start recounting an incident actually experienced 5 years ago (say) as something that happened then. The therapist needs to support the re-living as though it were happening NOW! The more detail the client can describe, the more real the experience, hence it is necessary to ask, "what can you see"? "What can you hear"? "What can you smell"? etc. This is painstaking work and needs to be done gently, supportively and at the pace of the client. Obviously, given the nature of the traumatic events, it is often not possible even if it were desirable to expose the client to real-life exposure for the purposes of therapy. The 'next-best' option is therefore re-living. If ever any situation puts me in awe of the bravery of clients it is this. To trust a therapist with the details of such distressing experiences and feelings is one thing – to deliberately expose oneself to those feelings all over again for therapy is another level of brave.

Once the client understands the process, rationale and – this must be emphasised – that they are in control of when to stop, it is time to begin. The sessions are recorded. The point of this is so the client can replay the recordings at regular intervals to re-experience the re-living as it were. When they re-play it, they are asked to rate their anxiety on a scale. As they continue their exposure, in theory, their anxiety levels will reduce in the same way as happens when exposed to other sources of anxiety. It is the fear of the event which leads to avoidance which maintains the problem. This is no different to other anxiety disorders from this point of view – though the treatment process is, from necessity, markedly different.

Clive suggested he play his recording in the car driving home – I advised against this. It is important that the client be able to complete all parts of this therapy in a safe place, with no disturbances where they can concentrate fully on the re-living experience and exposure. Again, the therapist needs to be alert to signs of avoidance, intentional or otherwise.

After four weeks from start to finish, Clive's flashbacks disappeared and he reported much lower levels of anxiety – even when the helicopters made an appearance. It is likely that hearing helicopters will always have a slight effect on Clive's pulse and anxiety – it would be a rare person who experiences what he did to no ill-effect. However, his quality of life is much improved and he is no longer feeling the need to duck into a pub – at least not for feelings of panic!

One of the fascinations of OCD, in common with many other disorders, is that some individuals experience all the symptoms while others, who may have been through ostensibly near-identical situations, experience nothing or little more than a transient anxiety or depression. This is, for many of us, what makes working in mental health so intriguing. Of course, for behaviourists, much of this is down to how we interpret events. Two people sharing the same rollercoaster car may feel completely differently about the experience – one views the ride as terrific fun while the other is convinced they are going to die (Grant *et al.* 2008). But that doesn't seem to tell the whole story – especially in a war situation. It would be odd to the point of pathological to feel anything other than horror at being in danger for one's life and/or seeing companions blown up and dismembered. Clearly there is something else here.

Much has been written about catastrophic misinterpretation (Gellatly & Beck, 2016) and the role it plays in panic. The usual explanation is that it is due to a complex mixture of factors – often genetic and early-life experiences but there is clearly much we have yet to learn about why some people misinterpret danger signs more readily than others. The truth is, we seem better at treating it that we do understanding it. If this seems a little 'cart before the horse' - I agree. It would be ideal to know how this occurs in order to create a more scientific formulation. However, for now, success in treatment will have to suffice.

Izzy – Perfectionism

As the saying goes, "I wish I had a pound for every…" – in my case for everyone who states with pride that either they or someone they know is a perfectionist. It seems to be worn as a badge of honour by many. The notion that one has very high standards, isn't easily pleased, won't stop until they are completely satisfied sounds reasonable when taken at face value. The worker who is painstaking in their dedication to detail, the student who won't settle for less than an 'A' grade, the athlete who must always win – preferably by setting a new world record, the diner who demands the highest standards of service, this 'type' is celebrated in popular media as something to aspire to – "because you're worth it".

There is no doubt that the opposite of perfectionism suggests a rather 'couldn't care less' approach, which, if one is a customer or employer, doesn't sound so appealing. Who wants their dinner cooked by a lazy chef or their heart operation performed by a less than dedicated surgeon? The problem is, perfectionism doesn't distinguish between the seemingly trivial and the vitally important. It can place the same importance on painting a wall as saving a life. In this, it can mimic OCD in many ways but the roots are often different (Wells *et al.* 1995).

OCD often involves fears of something bad happening if I don't do 'X'. Therapists generally agree that at its core, beliefs around fear of uncertainty and the desire to maintain a feeling of control over that uncertainty are present (Wells *et al.* 1995).

Perfectionism, however, often reveals at its core feelings of profound personal inadequacy. A feeling that personal worth is often reflected in one's deeds – a form of status anxiety if you will. Popular aphorisms abound on the subject – "If a job is worth doing…" etc. Early conditioning is found to play a large part in this (Flett *et al.* 2016) – most perfectionists can trace their traits back to clearly remembered messages from (usually) parents or one parent in particular. I would never claim to be a perfectionist – I'm probably too far the other way in some respects, but I can still remember my father stating quite categorically that once a job is started, it <u>must always </u>be finished. This sounds fair enough – we all know people who have huge numbers of unfinished DIY projects etc. throughout the house, or books that they start reading that they never seem to get time to get to the end of but again, it is knowing which is vital and which isn't. He was trying to teach me staying power, of course. An important life lesson and one for which I have had cause to be grateful on numerous occasions. The times I nearly dropped out of nurse training, for example, because I was hating the 3-month practice placement or unhappy with the grades for an essay, but if I hadn't stuck it out, I wouldn't be writing this now – I would have missed out on so much.

But balance is key. There is no point in watching a film or reading a book or finishing a meal if you are deriving no pleasure from it. Yes, it might get better later but why invest time in something so unimportant if the rewards, if any are so slight? Of course, if it is a job or a relationship or a course of study which will change your life then the consequences of opting out are potentially much more serious – it's about proportionality. The trick is to know which things are worth it and which aren't – my father's advice, though well-meant, left out this important detail and it took me years to work it out for myself.

With perfectionism, the nuances outlined above are similarly missing. Perfectionists find compromise in their high standards in specific areas very challenging. To many, 'good-enough' is ok. 'Good -enough' can be ok to the perfectionist too except for certain situations – someone who derives much self-esteem and pride from their garden will spend hours getting the lawn just so, the plants regimented and colour co-ordinated and the hedge symmetrical. They may not be so fastidious in how they dress or what they eat. For other perfectionists, the trait is much more of a common feature in their lives. We probably all know someone who will "not settle for second-best" and will only drive a premium car, wear expensive designer clothes and want top of the range gadgets. This tends to be more about status-anxiety than perfectionism – a feeling that one's worth (they are often not consciously aware that they are doing this) is projected outwards as someone who is **someone** if they wear the "right" perfume or t-shirt. Only a "loser" drives the "wrong" car. Perfectionism tends to be slightly different (Flett *et al.* 2016). People with status anxiety tend to constantly compare themselves with others and have an overwhelming need to feel superior. Sufferers of perfectionism tend to beat themselves up to such a degree over getting it "right", they don't notice others so much. Do you recognise the situation of painting a ceiling only to sit back after I have cleaned up and notice the bit I've missed – then, have to get all the paint and brushes out again over and over? I remember a man who used to build model sailing ships with match sticks. He would spend months, painstakingly building the ships in incredible detail, trimming the matches to size, carefully gluing them together only to be dissatisfied in the end and dismantle them to start again because it wasn't quite right to him (to others, it looked very impressive). I also remember nursing a middle-aged woman in day hospital in the 1990s who was a self-confessed perfectionist. She was most unhappy unless the house was pristine – every newspaper tidied away, every cushion in its place, no dust anywhere. The poor woman was exhausted as she was constantly cleaning. She had assumed this was normal. She had internalised early parental lessons such as "only dirty women leave the house with washing-up in the sink" and "a good wife makes sure the house is always spotless for her husband" (feminists reeling everywhere at this!)

"I bet your house is spotless" she declared to me once, out of the blue and was genuinely puzzled by my amusement at her remark. This lady was shocked to hear me and some other staff saying our houses only got cleaned about once a fortnight – we were too busy or too tired to do it more often (and frankly didn't care about a bit of dust either). I remember her saying "you're joking, I don't believe you, don't be silly" to this – she really thought cleaning the house very day was normal.

Now, some of you will be thinking "ooh, I could do with a cleaner like her" but rally, it would do you no good because the poor woman wouldn't have time to clean your house as she spent eight hours a day cleaning here own – every day! Perfectionists are made miserable by their obsession.

But I digress – in the case above, the perfectionism was obvious and easy to help. In Izzy's case, it was much more of a challenge first to identify and then to treat. In common with many mental health difficulties, there were easily identifiable and common thinking distortions, such as living by fixed rules and all or nothing thinking (Beck *et al.* 1987).

Izzy was referred via her mother, a district nurse who had attended one of my days teaching at the university. I had been asked to cover common mental health problems and their treatments in a morning – as you do – and a few days later, was surprised to receive an email asking if I did private work. It seemed her daughter, Izzy, had been suffering with anxiety for some time and it was affecting her work. On meeting Izzy, she seemed like any other lively, intelligent, attractive young woman in her early 20s. Izzy had qualified as a school teacher but towards the end of her teacher training had started to experience severe panic when working on placement. It was affecting her sleep and her overall mood. She was starting to consider the possibility that she had trained as a teacher only to find she couldn't face teaching. It seemed at first to be a classic case of panic disorder. There was the easily recalled first episode of panic, followed by the avoidance and the worry which became a preoccupation and further avoidance. Izzy was now a qualified teacher but had applied for a post as a basic grade teaching assistant. This she enjoyed and was coping well. It was clear that she felt more comfortable because of the reduced level of responsibility. This was a primary school for children with behavioural and learning difficulties and it was clearly a very challenging place to work. Izzy had a high degree of self-awareness and knew she was avoiding – a fact she felt very guilty about and she became tearful as she recounted her feelings of failure "all that training and hard work gone to waste" – understandably, Izzy was very disappointed and not a little angry with herself.

It didn't take too long to ascertain that her feelings of panic coincided with the stage in her training where she was having to take more and more responsibility for classroom management. This was a taste of her professional future and she found it very difficult to work with. When a client discloses information relevant to the formulation of their current situation (Grant *et al.* 2008), it is important that the nurse/therapist resists the temptation to cut corners. It is very easy to take snippets of information and put two and two together to make five – usually because it fits a preconceived theory or mirrors something in your own mind or as a result of some other confirmation bias (Parmley, 2006). Patience is needed – patience, restraint and the ability to elicit further information. Not some Hercule Poirot moment of clarity but some semblance of gentle probing. I asked if the idea of being in charge might be something that worried Izzy? She was ahead of me, she said it did. The next question could have been "Why"? However, I always advise against "Why?" questions. If she knew why, she probably wouldn't be here. Why will only get you so far, anyway. It is important to note here that two things made this process a lot easier:

1) Izzy was bright and self-aware. She had thought about this a lot and was half-way to sorting things out for herself. Like anyone else, I believe she just needed some support – an empathetic ear and some ideas to help further.

2) My background in education. OK, I'm not a primary school teacher and never could be but I do have some idea of the pressures she faced. She knew this (see earlier chapter on PTSD and service personnel) and this helped form the therapeutic relationship as I had higher credibility to her (Henretty *et al.* 2014).

I needed to know what it meant <u>for Izzy</u> to be in a position of "not coping". I asked what was the worst that would happen. Izzy said she didn't know so I re-phrased the question, would anyone come to any harm? She replied probably not. I asked Izzy to imagine a scenario where her worst fears came true – what would she do? She said she would ask for help from another teacher.

Therefore, we established two things:
1) No-one would die
2) The situation could be resolved

However, this didn't solve the problem – we needed to go further. Building on the above scenario, I asked Izzy what it would say about her. She said it would make her look weak, ineffectual, a bad teacher (tears again).

I asked if she would think these things if it were someone else in her shoes (this is a high-risk question – if she says yes, I've learned something else about her but backed myself into a corner).

Thankfully she said no. She just felt that the class should be run well. Obviously, I needed to know what she meant by this. Izzy said a well-run class should (that word again cf.

musterbation, Ellis) almost run itself with a minimum of disruption and that it was her responsibility.

Most of us would agree that this sounds like the ideal. Most of us probably have memories of at least one incident at school where the class was disrupted by one or more unruly pupils and the response of the teacher, which in the case of my school days varied from completely hapless to aggressive, over-the-top authoritarian.

However, to return to the notion of good enough, not all classes can be completely smoothly-run all of the time. That, I proposed, was simply unrealistic especially in a school with such demanding pupils – with the best of wills, that was simply expecting too much of herself and the class.

I asked Izzy if other teachers had 100% success in this area? She replied she didn't think so, so why was she being so hard on herself? Izzy couldn't answer this as she had been so consumed by the sleepless nights of worry about the next day. It transpired that the perfectionism stemmed from her carefully planning each class of the day as she had been taught at university, in minute detail. Lesson plans are very helpful tools. They divide the session into manageable chunks of content and ensure what is needed to be taught is taught, with breaks and different teaching and learning strategies built-in, but that is all they are, tools. What was so distressing Izzy, was when she had to deviate from that plan because of the needs of an individual, who might disrupt the class and take time to calm down. Now I was understanding the problem more and so was Izzy. Through taking time and using Socratic questioning and downward arrow technique (Grant *et al.* 2008), we had gone from fear of looking like a poor teacher to fear of having to deviate from a carefully constructed lesson plan. The consequences of varying from the plan was something Izzy couldn't easily identify but it generally came back to looking "foolish" and of the children not learning what they needed to.

The first part of this was the easy one to address. Planning for contingencies was something I was well used to in my time as a tutor and we soon managed to design teaching plans to take this into account. The second part took a little more organizing. We needed to create a situation where Izzy relaxed some of her tightly-held control in a way which was safe for the children but real enough to have some learning value for her. Izzy was a teaching assistant but also a qualified teacher. I asked if she ever ran a class on her own – that was her responsibility, if only for a couple of hours or so. She said she did, but she planned it with the responsible teacher. I asked if it were possible for her to run such a class but only plan for the first half and leave the rest to chance?

Those of us who have been in teaching for a while are quite comfortable with doing this. It comes with confidence and knowledge of our subject, our ability to improvise and our students. Izzy was working in an arguably more demanding environment where the children could be unpredictable and needed more monitoring and attention from the staff.

Izzy was able to discuss this with the teacher with whom she got on very well and a plan was hatched. The rest of the story is that this was a success. Of course, we prepared for the session with Izzy writing down what her worries would be and her responses to read to herself the night before. I also asked her to leave these downstairs after she had written them, so that she could sleep – a well-known technique (Akerstedt, 2006).

A week or so after the session, we debriefed. What had gone well? What had her predictions been? What had happened? Did the predictions come true? What had she learned? The answers to these questions all led me to feel things had gone well. Izzy looked and sounded much calmer. She remarked that others had commented on how much more relaxed and happy she looked. She had come a long way and although there were obviously going to be times in the future when she would face related and unrelated challenges, she could be proud of taking a big step forwards in her coping abilities and teaching skills repertoire.

As stated previously, it is vital in CBT that the skills learned through practice are not just completed once but regularly – use it or lose it. Izzy continues to take responsibility, only planning 75% of the session and leaving the rest for however the mood (of herself and the class) takes her and she is enjoying her work. She plans now to apply for a full teaching post at a mainstream school and finally work at the level for which she has trained. She has suffered something of a hard apprenticeship, through her difficulties but also because this would be a tough place to work for any teacher. However, she knows that the skills she has gained, partly through therapy but mainly through her own willingness to learn and respond to challenges are going to provide her with many of the skills she will need for the future.

Notes on Notes

1) The NMC Code

One issue which employers, managers but especially practitioners need to bear in mind is the NMC Code of Professional Standards of Practice and Behaviour for Nurses and Midwives (NMC. 2015). An often poorly-understood and even worse, poorly-quoted document, like the European Working Time Directive (www.ec.europa.eu/legal-content 2003) its purpose is to protect not just the public but the individual practitioner from the employer who asks more than the worker feels able to give. In the case of nursing, the NMC Code Section 13.3 requires the nurse to;

"Ask for help from a suitable qualified and experienced healthcare professional to carry out any action or procedure that is beyond the limits of your competence" (NMC 2017)

Alongside this, we see, in Section 13.5;

"Complete the necessary training before carrying out a new role".

Clearly, although these statements are to protect the public, they are also to protect the nurse. I am not a phlebotomist. I have had no training in phlebotomy. If I were asked to take blood, I would refuse, citing if necessary the above Sections of the NMC Code. I know I would make a pig's ear of taking any poor soul's blood so would not even try. I could ask to be trained but it is not something I particularly want to do, if I'm being honest.

However, I would and do argue, that most if not all of the skills I have demonstrated above, could be carried out by a competent mental health nurse. Perhaps there is a case for including much more in the way of CBT skills in the nursing curriculum and this may well be coming in the revised standards for nurse training and education (NMC. 2017) – there are hints of this in the coming years.

I would also strongly recommend that supervision from a suitably qualified and accredited CBT therapist is obtained if the nurse is hoping to use all but the most rudimentary CBT skills. This is not to downplay the skills of the nurse – even accredited therapists are required to get clinical supervision from appropriately qualified and experienced colleagues. Who pays for the training and supervision is of course, sometimes a moot point. However a case can be made for the financial efficiency of such practice, even with such 'extras' in place.

One final note on this – the Code also states that the nurse should;

"Raise your concerns immediately if you are being asked to practise beyond your role, experience and training". (NMC. 2017 S.16.2)

This is to reassure those of you who are not at a stage where you feel ready to work at this level, whether through lack of training, supervision or volition – though I would refer you back to the Revised Standards (above) – the day is coming – soon!

Current clinical colleagues tell me that some mental health trusts are encouraging nurses to attend stand-alone courses, usually 10 days to learn the fundamentals of CBT. They then document interventions stating that "a CBT style" was used as an intervention. This seems to be professionally satisfactory.

2) Treading on toes

The history of healthcare is packed with professional groups all jealously guarding their own specialities. Nurses today do much of what was the sole preserve of doctors not very many years ago (prescribing, phlebotomy, scans, clinics, 1:1 therapy, minor surgery etc. etc.). In 2006, the Layard Report (cep.lse.uk/ 2006) identified the need for 10,000 extra therapists to provide psychological therapies. This was the birth of IAPT (Improving Access to Psychological Therapies). Cynics suggested this was a cheap way to provide more support to get service users with anxiety and depression back to work. Therapists were trained quickly and to a level which lacked the depth and intensity of the more traditional routes (I have taught on both programmes so am well-placed to know this). The therapy provided is, no doubt, of great benefit to many, many people. IAPT therapists are often sited in GP surgeries and so access is indeed improved – physically. However, demand still outstrips supply (Independent, September 15[th]2014) (though on-line help is more readily accessible) and ironically, many in-patient areas have no such access as for many years now, the emphasis has understandably been on treating people on an outpatient basis. As an inpatient, one would need to be discharged in order to access further help, which may still take several months to obtain.

Therefore it surely makes sense for inpatient areas to have access to practitioners with the skills to at least start this process. It could potentially free-up beds more quickly at a time when they seem to be at even more of a premium than ever before and reduce the need to prolong suffering.

3) In Conclusion

This snapshot of some of the people I have worked with in the last 26 years represents some of the more memorable cases. I could have chosen many more – the man made redundant for whom nothing seemed to improve his mood until he was given ECT (it wasn't within our power to get him a job, alas) the young woman sexually abused by her stepfather who had flashbacks, the woman who showed no suicidal ideation whatsoever who walked in front of a train – they were all memorable too and maybe, just maybe I'll get around to putting them into the next book – if there is one. But the point I wanted to make was that all of these people are both remarkable and unremarkable. Remarkable because they chose to seek help and chose to trust a complete stranger with their innermost fears. It never ceases to amaze me, even now, that people can come to trust when they need to, someone whom they have never met with such personal information. It really does humble me to think that this happens – I ask myself if I could do this and I'm really not sure – I would need to be very motivated and it is true that they were.

Unremarkable though, because to pass them in the street, to talk to them on a bus, there would be nothing in particular that would make them stand out – except of course they were all pleasant, polite people.

This is significant because I wanted to show that mental illness doesn't just happen to other people, it happens to us all and the sooner we understand this fact – really understand it – the sooner we can get it taken seriously by those who hold the purse strings and tell us they care when they clearly don't.

References

1. Akerstedt, T. (2006) *Psychological Stress and Improved Sleep.* Scandinavian Journal of Work, Environment and Health. 493-501

2. Andressen, R. Oades, L. Caputi, P. (2011) *Psychological Recovery: Beyond Mental Illness.* Wiley-Blackwell

3. BABCP (2017) *www.babcp.com* accessed 12.09.2017

4. Barker, P. (2008) *Psychiatric and Mental Health Nursing: The Craft of Caring. 2nd Edition.* Taylor & Francis

5. Beck, A. T. Rush, A. J. Shaw, B. F. Emery, G. (1987) *Cognitive Therapy of Depression* New York. Guildford.

6. Bowyer, L. Wallis, J. Lee, D. (2014) *Developing a Compassionate Mind to Enhance Trauma-Focused CBT with an Adolescent Female: A Case Study.* Behavioural & Cognitive Psychotherapy 42.2 248-254

7. Bream, V. Challcombe, F. Palmer, A. Salkovskis, P. (2017) *Cognitive Behavioural Therapy for Obsessive-Compulsive Disorder.* Oxford University Press

8. Clark, D.M. Hackman, A. Salkovskis, P. (1999) *Brief Cognitive Therapy for Panic Disorder: A Randomised CT.* Journal of Consulting & Clinical Psychology 67.4 583

9.   Fisher, C.B. Oramsky, M. (2008) _Informed Consent to Psychotherapy: Protecting the Dignity and Respecting the Autonomy of Patients._ Journal of Clinical Psychology 64.5 576-588

10.   Fisher. L. B. Sprich, S. E. (2016) _The Massachusetts General Hospital Handbook of Cognitive Behavioural Therapy_ New York. Springer.

11.   Flett, G. L. Hewitt, P. L. Sherry, S. S. (2016) _Deep, Dark and Dysfunctional: The Destructiveness of Interpersonal Perfectionism._ APA

12.   Forster, S. Bennett, R. (2001) _The Role of the Mental Health Nurse (Mental Health Nursing & the Community)_

13.   Gellatly, R. Beck, A. T. (2016) _Catastrophic Thinking: A Transdiagnostic Process across Psychiatric Disorders._ Cognitive Therapy & Research 40.4 441-452

14.   Gibbons, B. Birks, M. (2016) _Is it Time to Re-Visit Stigma? A Critical Review of Goffman 50 Years-On_ British Journal of Mental Health Nursing 5 (4) July/August 2016 pp 185-189

15.   Gilbert, P. (2000) _The Relationship of Shame, Social Anxiety and Depression: The Role of the Evaluation of Social Rank_ Clinical Psychology and Psychotherapy 7, 174-189

16.   Gilbert, P. (2007) _Psychotherapy and Counselling for Depression_ London Sage

17.   Grant, D. Mills, J. Mulhern, R. Short, N. (2006) _Cognitive Behavioural Therapy in Mental Health Care_ London. Sage

18. Grant, A. Townend, M. Mills, J. Cockx, A. (2008) *Assessment and Case Formulation in Cognitive Behavioural Therapy* London. Sage

19. Greenburger, D., C Padesky. (2015) *Mind over Mood: Change the way you feel by changing the way you think 2nd edition.* New York. Guildford.

20. Henretty, J. R. Currier, J. M. Berman, J. S. Levitt, H. M. (2014) *The Impact of Counselor Self-Disclosure on Clients: A Meta-Analytic Review of Experimental and Quasi-Experimental Research* Journal of Counselling Psychology 10. 1037

21. Independent (Editorial) September 15th 2014

22. Kuyken, W. Padesky, C. Dudley, R. (2008) *The Science and Practice of Case Conceptualization* Behavioural and Cognitive Psychotherapy 36.6 757-768

23. Leahy, R. Holland, S. J. (2000) *Treatment Plans and Interventions for Depression and Anxiety Disorders* London. Guildford

24. Mittal, C. Griskavicious, V. (2014) *Sense of Control Under Uncertainty Depends of People's Childhood Environment: A Life History Theory Approach* Journal of Personality and Social Psychology 107.4 621

25. Moghaddam, N. G. Dawson, D. L. (2016) *Cognitive Behavioural Therapy* e-prints. University of Lincoln. Accessed 12/09/2017

26. NICE Guidelines www.nice.org accessed 12/09/2017

27. Nolan, P. (2000) *A History of Mental Health Nursing* Thomas Nelson Publishers

28. Norman, I. Ryrie, I. (eds) (2013) *The Art & Science of Mental Health Nursing: Principles & Practice* McGraw Hill.

29. Norton, T. R. Manne, S. L. Rubin, S. Hernandez, E. (2005) *Cancer Patients' Psychological Distress: The Role of Physical Impairment, Perceived Unsupportive Family and Friend Behaviours* Health Psychology 24.2 143

30. Parmley, M. C. (2006) *The Effects of Confirmation Bias on Clinical Decision-Making* Drexel University

31. Phillips-Sheesley, A. Pfeifer, M. Barrish, B. (2016) *Comedic Improvisation for the Treatment of Social Anxiety Disorder* Journal for Creativity in Mental Health 11.2 157-169

32. Rauch, M. Sheila, A. Eftekhari, A. (2012) *Journal of Rehabilitation Research and Development* 49.5

33. Rogers, A. Pilgrim, D. (2005) *A Sociology of Mental Health & Illness* McGraw Hill

34. Tambuyzer, E. Pieters, G. Van Audenhove, C. (2011) *Patient Involvement in Mental Health Care: One Size Does Not Fit All* Health Expectations 17. 138-150

35. Taylor, S. (2017) *Clinicians' Guide to Post Traumatic Stress Disorder* London. Guildford

36. Thomas, M. Drake, M. (2012) *Cognitive Behaviour Therapy Case Studies* London Sage

37. Weatherhead, S. Flaherty-Jones, G. (2012) *The Pocket Guide to Therapy: The 'How To' of the Core Models* London Sage

38. Wells, A. Clarke, D. M. Salkovskis, P. Ludgate, J. Hackman, A. Gelder, M. (1995) *Social Phobia: The Role of In-Situation Safety Behaviours in Maintaining Anxiety and Negative Beliefs* Behaviour Therapy 26.1 153-161

39. Westbrook, D. Kennedy, H. Kirk, J. (2007) *An Introduction to Cognitive Behaviour Therapy: Skills and Applications* London Sage

40. Willis, F. (2008) *Skills in Cognitive Behaviour Counselling & Psychotherapy* London. Sage

41. www.cep.lse.ac.uk/ accessed on 19/09/2017

42. www.ec.europa.eu/legal-content accessed on 19/09/2017

43. www.gov.uk/ukds (2010) *Equality Act* Accessed 17/09/2017

44. www.nice.org accessed 17/09/2017

45. www.nmc.org/standards/code accessed 17/09/2017